I0487470

Inner Algebra

How To Do Algebra In Your Head

Aaron Maxwell

Published by Hilomath (`http://hilomath.com`).

This book was set in the Times font using LaTeX, and written with the LyX editor (`http://lyx.org`). The figures and cover were assembled using The GIMP (`http://gimp.org`). The math symbols in the figures were created with the L2P tool (`http://redsymbol.net/software/l2p/`).

We like to hear from you. For publisher correspondence, please send email to service@hilomath.com, or postal mail to Hilomath, Box 225120, San Francisco, CA 94122, USA.

Contents

Acknowledgements

Thanks to my friends Colby Lemon and John Chodera, and my sister Annette Maxwell, for encouragement and feedback. Thanks to my longtime friend, Scott Tabbert, for his enthusiastic support, despite him not being a math person *at all*. And thanks to my housemate Kate McLaughlin-Williamson for valuable suggestions on the cover design. (Be glad I asked her for feedback. If you can stand to look at it for very long, you have her to thank.) These people only helped improve it, of course – if anything is less than ideal, that is my doing.

This book has been written entirely using open source software. The platform was a Debian GNU/Linux system, using either the icewm or the KDE window manager environments. Some other software used includes LyX and the LaTeX typesetting system, The Gimp (for image creation and editing), the ImageMagick image processing tools, Perl (for making various support software), Subversion for document version control, the use-for-all-sorts-of-things tools such as emacs, vi, konsole, konqueror... and many others.

All of this software exists because, worldwide, developers I never met invested their time and energy to write software, with the purpose of freely giving it to anyone who wants to use it. This book simply would not exist in this form were it not for those developers' generosity. My tremendous thanks and appreciation to all of you.

Important Note for Students

1. If your instructor asks you to write out all the steps you take to solve an exercise, please do that... even if (because you read this book) you can solve the whole thing in your head.

2. For best results, only use the methods in this book to do math that you can already do "normally" (by writing it out). If you have trouble doing a certain calculation on paper, wait until you CAN do it that way before using the methods in this book to solve it.

If you are in school and taking a math course right now, your instructor will probably want you to "show your work" on homework and tests. This book will teach you how to correctly do those exercises while showing less of your work. If you practice what's in here, you will probably be able to solve some exercises entirely in your head. That's really great, except that your teacher wants you to write out intermediate steps.

Please make sure you actually do write out everything the way your instructors would like. Your instructors are responsible for making sure you get a certain level of math knowledge (assuming you do your part); that's their job. Writing out everything helps them do that. More importantly, it also helps you learn the math deeply. Of course, you are the one who benefits from all this.

This book assumes you already know algebra, and can use it to solve equations. If you are still learning algebra, that's okay – you can still benefit from this book. The only requirement is that you have some experience solving algebra equations. As you read, if there are parts that talk about math that you haven't learned yet, just skip over those sections for now.

There is a difference between mathematical knowledge, and what goes on inside of you when you apply that knowledge. This book is only about the second part. There is a bit of trickiness here. Some of the techniques in this book are, in a sense, shortcuts. You can use them to mentally bypass steps in a calculation. But if you use them to do math that you don't yet understand how to do normally, then it's possible that you will apply the techniques incorrectly. You will still get an answer, but it will probably not be mathematically correct. Even worse, you may "learn" faulty math habits.

Fortunately, there is an easy way to handle this: only use the techniques in this book to do math that you already know how to solve normally (on paper). If you are not sure, attempt to solve it on paper first. As you do this, you will get better and better at knowing when it is safe to use the techniques in this book. Following this guideline will allow you to safely get the greatest long-term value from this book.

Preface

Many mathematicians and engineers can look at an equation, think a moment, and know its solution. Some can do very complex math this way. Most people can learn to do this. By using certain mental abilities you have in ways described in this book, you can do things mathematically that may have not seemed possible before.

In this book, we focus on algebra; you learn to solve algebra equations mentally. It is necessary that you have already learned algebra in some way (probably through a math course). What you learn here is to do certain things that allow you to do algebra internally. If you are just learning algebra (for instance, if you are now taking a course in school), you will begin to find this book useful as soon as you have some experience solving equations. (You will mainly be working with what are called *singular* equations, which are explained below. For now you'll ignore the other kind, called *plural* equations.)

The skills taught in this book come in several stages. In the first, you learn to solve algebra equations much faster. For simple problems, this means you will know the solution just by looking at it and thinking a moment. If the equation is more complex, you will solve it with fewer written steps than before. This first stage will take place after you read through the end of chapter 3, working the exercises along the way.

As mentioned, we can categorize algebra equations as *singular* and *plural*. An equation is usually singular if it can be reduced to the form "x = <some number>". A singular equation has one variable and a single solution. Here are some singular equations:

$$x + 2 = 3$$

$$\frac{x - 2.7}{42} = \frac{2\pi}{3}$$

$$2 = \frac{x + 2}{x - 5}$$

$$\sqrt{x+1} = 3.9$$

An equation is *plural* if it is not singular. There are several kinds. Examples of plural equations are those with more than one solution, like quadratic or higher-order polynomial equations; or equations like $\sin x = 0.7$.

(By the way, if you are wondering why the first type of equation is called *singular* and the second type is called *plural*, it has to do with how you will learn to solve them. This is explained more fully in chapter 6.)

In the second stage, you start to solve singular equations without writing intermediate steps. You will look at it, do some things mentally, and know the solution. You get to the second stage by practicing what you learn in stage 1. If your job or school work involves algebra, getting this practice is simple. Since you'll be manipulating equations anyway, you just practice these techniques while doing it. As you practice, you get better, and naturally need to write fewer steps down. (It may or may not be *easy* – since you are relearning and reshaping habits, it will take some dedication on your part.) There will also be equations you can solve by just looking at them and thinking for a moment, whereas before you started this book, you would have needed to write a lot to solve it.

In the third stage, you will start to solve plural equations in your head. These equations have some special qualities. The skills for solving them build greatly on those for solving singular equations.[1]

There is a web page for this book (`http://inneralgebra.com/help/`). Any additions, corrections, FAQS, or other helpful information will be put on this page. Also, feel free to contact me by email if you have questions or comments (amax@hilo-math.com).

---

[1] Some plural equations are easier to solve than others. Quadratic equations are plural equations, and are relatively easy to solve. You will be able to get them soon after you gain some experience working singular equations.

Chapter 1

Quick Start

We are going to start off with a demonstration. We will take an equation and explain, step by step, how you would solve it mentally. As this happens, you will see many of the techniques and concepts that can be used on other algebra equations. You will be able to apply some of it immediately after reading this chapter.

Here is the equation we will solve.

$$0.5 + \frac{4x + 3}{2} = 1.5 + x$$

Depending on your experience level, this may be challenging, or it could be really easy. In any case, you could solve this by first subtracting 0.5 from each side,

$$0.5 + \frac{4x + 3}{2} = 1.5 + x$$
$$-0.5 \qquad\qquad -0.5$$

$$\frac{4x + 3}{2} = 1 + x$$

then multiplying each side by two,

$$2 \times \left(\frac{4x + 3}{2}\right) = 2 \times (1 + x)$$

$$4x + 3 = 2 + 2x$$

and then collecting the variable by itself on one side.

$$4x + 3 \;=\; 2 + 2x$$
$$-3 - 2x \qquad -3 - 2x$$

$$4x + 3 - 3 - 2x \;=\; 2 + 2x - 3 - 2x$$

$$2x = -1$$

$$x = -\frac{1}{2}$$

Most people would work this out on paper (or a whiteboard or chalkboard), writing some of the steps that were printed above. Math as we know it is primarily *visual*: equations and symbols are represented as something that can be seen. You read or write an equation, and probably do not hum it or smell it. The methods we have for solving equations reflect this. One such "tool" is subtracting a quantity from each side. We used this when we took the starting equation

$$0.5 + \frac{4x + 3}{2} = 1.5 + x$$

and changed it to

$$\frac{4x + 3}{2} = -0.5 + 1.5 + x$$

What we are doing is removing a term from one side, in a way that moves us towards the solution, while keeping the equation mathematically correct. We represent this visually – when writing it, or printing it on a page – by printing "-0.5" under each side of the equation.

$$0.5 + \frac{4x + 3}{2} = 1.5 + x$$
$$-0.5 \qquad\qquad -0.5$$

Let's talk about something else a moment. Like most people, you have certain mental abilities, including a memory, an imagination, and the ability to visualize in your mind's eye. Some people are much better at this than others, but most everyone has some innate skill in these areas. It is also something that people can learn and become better at.

You can use these abilities to solve equations mentally. That, in a nutshell, is what this book is about. I'd like to demonstrate what this means, so I am going to ask you to do something. Please take a moment to close your eyes, and see if you can visualize this equation:

$$0.5 + \frac{4x+3}{2} = 1.5 + x$$

(Well... read the equation first, THEN close your eyes.) Just try it for a few seconds – if it does not come quickly, don't worry about it. You are seeing the equation in your mind's eye. If it is not easy to do, try just visualizing part of it – say, the $1.5 + x$ on the right side. Were you able to do this? This isn't a small equation, so maybe you did not. It's really not important at this point whether you can, because part of what this book does is teach you how to do that. For now, just pretend you have the ability to see equations like this using your imagination – just like an artist can visualize what he will draw before picking up his pen.

If you have the ability to visualize this well, you can do things mentally that you would normally do on paper. To move the 0.5 to the opposite side, you could start by seeing something like this:

$$0.5 + \frac{4x+3}{2} = 1.5 + x$$
$$-0.5 \qquad\qquad -0.5$$

This is similar to what you do if you are working it out on paper, except that you are *imagining* it rather than writing it. You would then go on to actually do the subtraction, and so on. That's great, but there is something else we could take advantage of here. The procedure above – subtracting a quantity from each side – is presented visually in a certain way. Part of why it is the way it is, is so that it can clearly be represented on paper (or whiteboard, computer screen, etc.) Since we are working within an image that you are visualizing, there are things you can do that could not really be done in any of those mediums. The immediate goal is to take the 0.5 term and remove it from the left side of the equation. Let's see if there is a different way we can accomplish this. Here is the original equation again, below. See it in your mind's eye, if it is easy for you do to so right now. If not, just pretend that is what you are doing as we go along.

$$\boxed{+0.5} + \frac{4x+3}{2} = 1.5 + x$$

A few things are different here. For one, we have put a plus sign in front of the 0.5. That's fine; it is a positive number anyway, so we are just emphasizing that fact. More unusual is that there is a box of dots surrounding the +0.5. That is a *chunk*, which is a visual device we use in this book. We say that we have *chunked* the +0.5. You can chunk a number, variable, expression – any group of symbols in an equation. You chunk an expression when you are getting ready to do something mathematically with it. In this case, we are moving the contents of the chunk to the opposite side of the equation:

$$+ \frac{4x+3}{2} = 1.5 + x$$

$$\boxed{+0.5} \rightarrow$$

If you are visualizing this equation, "pick up" the +0.5 in your image, and begin to move it, just like in the picture here. (If you are just pretending to visualize the equation right now, pretend you are moving the chunk too.) The chunk is like a separate object, that you are taking from its spot on the far left and moving to the right. As you move it, there is something special that you need to do with it. You need to take whatever is inside the chunk and multiply it by -1. Since the number is positive, we just make it negative, like so:

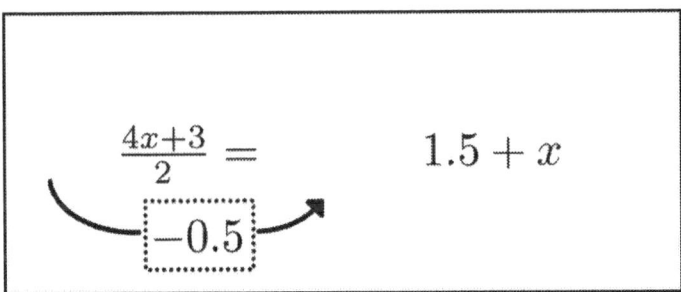

(We've also moved some of the symbols on the right side over, because we are about to insert the chunk there.) Whenever you move a chunk like this, you change the term's sign to its opposite, as if we are multiplying it by -1. If the term was negative – for example, say it was -0.3 inside the dots – we would make it positive ($+0.3$). The other thing we do here is to remove the $+$ that was between the $+0.5$ and the $\frac{4x+3}{2}$ when we first started.

The last step is to insert the chunk on the right side of the equation.

$$\frac{4x+3}{2} = \boxed{-0.5} + 1.5 + x$$

Notice that we are *adding* it in. We insert the chunk in a way that it is being added to the rest of what is on that side of the equation. We are done with this chunk, so we can remove the dotted box now.

$$\frac{4x+3}{2} = -0.5 + 1.5 + x$$

This is an example of *symbol motion*, which is fully explained in chapter 3. Symbol motion involves moving groups of symbols (chunks) in ways that algebraically manipulate the equation. When done correctly, it has the same mathematical effect as the normal methods. The difference is that once you learn to use them, symbol motions are easier and faster to use internally than the techniques used to solve equations by writing them.

After simplifying the right side a bit, the next step would be to multiply each side of the equation by 2. There is a way to do this with symbol motion also. It starts with chunking the 2 in the left side's denominator.

$$\frac{4x+3}{2} = 1 + x$$

Then we start to move it, like how we did earlier.

$$\frac{4x+3}{2} = 1 + x$$

There are several classes of symbol motion. When we moved the 0.5 to the right side earlier, that was one class, in which the effect is similar to adding or subtracting a quantity from each side. When we move the 2 from the denominator, we are using a different class that has an effect similar to multiplying (or dividing) each side by some quantity. Unlike with the "add/subtract" class, we do not multiply the contents of the chunk by -1. What we do instead is to move the chunk from the *denominator* of one side, to the *numerator* of the opposite side. (Or, if it was in the numerator of one side, we would move it to the other side's denominator.) In other words, if the chunk is on the bottom of one side, we move it to the top of the other side; and if it is on the top of one side, we move it to the bottom of the other side[1]. We end up with something like this:

$$4x + 3 = (1 + x) \times \boxed{2}$$

In the process of doing this, a few things in the image have changed again. First, there was a horizontal line in the fraction on the right hand side – now it's gone. Second, we introduced a multiplication sign on the right hand side, between the $1+x$ (which we put in parentheses) and the chunk. Symbols that denote some math operation, like $+$, $-$, $\times$ or $\div$ (which is what the horizontal line in a fraction really is), are special. They flit in and out of existence depending on what numbers or variables are present and their relationship with each other.

Let's finish solving this equation. Now that the 2 has been moved, you can dechunk it (remove the dots around it).

$$4x + 3 = (1 + x) \times 2$$

Next, in the visual image you have, you can distribute the multiplication on the right side. This is done just as it would if you were writing the equation out.

$$4x + 3 = 2 + 2x$$

---

[1]Keep in mind that any expression can be considered a fraction. Even just $x + 1$ can be written as $\frac{x+1}{1}$. $x + 1$ is the numerator, and 1 is the denominator.

Now we want to collect the variable by itself on one side. Chunk the $2x$ on the right,

$$4x + 3 = 2 + \boxed{2x}$$

and move it over.

$$\leftarrow \boxed{-\,2x}$$
$$4x + 3 = 2 +$$

Remember, when you move a chunk like this, you need to flip its sign (which is the same as multiplying it by -1). That it why it is now a $-2x$ instead of $2x$. We can actually insert it anywhere in the left side of the equation, just as if we were adding in a term, so let's put it by the $4x$.

$$4x + (-2x) + 3 = 2$$

The chunk is put in parentheses here to help make things clear. Since $4x + (-2x) = 4x - 2x$, when we dechunk the $-2x$ the equation becomes

$$4x - 2x + 3 = 2$$

After simplifying, we can move the 3 to the other side of the equation in a similar way to get

$$2x = 2 - 3$$

$$2x = -1$$

$$x = -\frac{1}{2}$$

For some of you who are reading this, the equation used above will seem complicated. For others, it will seem simple and maybe boring. The first group will be heartened, and the second group dismayed, to know that this book teaches by using example equations, most of which are simpler than the above. For the second group (and the first group too if you are interested), I want to let you know that although the equations used are "small" and simple, *everything* taught scales to math of almost any complexity. The concepts and tools are introduced using simple equations so that it is clear and obvious how to apply everywhere... including decidedly non-simple equations. The "challenge" – the word is in quotes, because it's not really hard – is to take the essence and apply it in new mathematical situations.

There is one last thing to mention. If you encounter an equation that is challenging for you to visualize – and no matter how good you get, there will **always** be equations that are bigger and badder – you don't have to envision the whole thing at once. There is a technique, called *windowing*, that lets you visualize only those parts of the equation needed to correctly do the next symbol motion. The practical effect is that you can solve equations that are larger than you can visualize right now. Windowing is detailed in chapter 4.

Chapter 2

Groundwork

There are a few key abilities that form a foundation for doing math intuitively. The first is visualization. It is similar to using your imagination. If you have always been a 'visual' person, you will find some things easier. The first part of this chapter helps you develop this ability. If you already feel skilled in this area, this section can help you strengthen it in the specific ways that help you use it for math. Some people are naturally quite good at visualizing. If you feel that you don't really need training in this area, you can skip the first section. Just make sure you can do the exercises at the end of the chapter.

The last section introduces a tool called *chunking*. Chunking is a way to group (and un-group) symbols and expressions in particular ways that help us solve the equation.

2.1   Visualization

Think of a character from a comic, TV show, or computer game that you like. Close your eyes and imagine this character – see them in your mind's eye. Even if it is a fuzzy or incomplete image, that's fine. Just get a definite visual impression of this character.

If you can do that, great! You already have the single most important ability that is necessary to do algebra in your head. You can skip to the next section, "Visualizing Equations," below.

If you could not conjure up this image just now, or you are not sure, don't worry. First, it is not necessary to be able to visualize in any great detail. Let's say the character you picked is a talking duck you liked to watch on TV when you were seven. If you can see just part of the face and bill for a split second, and then it's gone, you can visualize plenty well enough. So just focus on getting that.

Practicing a little goes a long way. Here are some tips:

- Try something. If it works, great. If not, try another approach. What helps one person visualize really well is often different from what will help another person.

- It will probably help to choose a character you have positive feelings about. Think of a small child watching a cartoon, and suddenly a character the child really likes comes on the screen. What does that child feel inside? Avoid characters that kept you awake all night.[1]

- Visualize with your eyes open. Visualize with your eyes closed. Which is more effective for you?

- If it's not easy to visualize the character you picked, pick a different character, or some object that has meaning to you. E.g., if you skateboard, imagine a skateboard, or last year's X Games champ flying on one.

- Find a real picture of the character, or draw one. Look at the picture. Close your eyes, and try to imagine that picture. Repeat. Use the picture as an aid.

- Bring different senses into it. Some people are "visual" people; they will usually be able to imagine things in their mind's eye immediately. If you are more of an auditory person, imagine this character talking or making a noise. If you are

---

[1]If they *still* keep you up, sorry for reminding you. Better sleep with the lights on tonight.

more of a kinesthetic, "feeling" person, imagine the character doing something physically active, and what they would
feel while doing it.

- Move the "physical" location of the image. When people
imagine something, they usually put it in a specific location
in the space around or inside them. Some people envision
it in front of them, or slightly up or to one side. It can also
be located inside your body, such as inside your head. Try
several locations to find what works best for you.

- Be willing to give it time. Work on it a little, and if it is not
easily coming to you, do something else. Try again in a few
hours or another day.

Practice until you can consistently get a mostly complete image,
and can hold it for a few seconds. When you can do that, continue
to the next section.

Visualizing Equations

Now we are going to try something with this equation:

$$x + 2 = 3$$

Since this equation is so simple, you probably know the answer
just by looking at it.[2] How easy or hard it is to solve does not matter
– we are going to do something else entirely with it.

Imagine that you have written this equation in big letters and
numbers on a 3 by 5 card, like this:

$$x + 2 = 3$$

---

[2]If not, you are reading a book that teaches you how, so don't worry.

Imagine that this card is floating in front of you, where you can see it. It's at a comfortable distance. You can move it around by imagining it. You want be able to see each individual symbol, 'x', '+', '2', '=' and '3', and their position in space relative to each other. Practice this until you can consistently see the equation on the card. It does not matter if you do it with your eyes open or shut. It can help to write the equation on a real 3x5 card, and glance at it as an aid.

By the way, you certainly do NOT need to discern it in great detail. If, in your imagination, you can recognize each of the five symbols, and get the order right, that's all that is necessary. In other words, as long as YOU can tell that the first symbol is an 'x', even if it does not really look anything like it, that's fine – you are the only one who will ever see it anyway!

Now, take the image and move it around. Move it a few inches left, right, up and down. (Your head is motionless. You are getting practice at controlling where you put the image that you see mentally.) Again, practice this until you can consistently do it.

Now make the image of the equation bigger and smaller. You are zooming in and out. Make it twice its original size. Make it half its original size. Practice until you can do it consistently.

Remember, in all of this, you're not *doing* anything with the equation, mathematically speaking. You are not taking any steps to solve it. You are merely creating a mental image of it, just like you could imagine a car or a banana or anything. You are then moving that image around while keeping the image more or less intact.

Here is a summary of the steps above:

1. Visualize the equation $x + 2 = 3$ as if it was written on a 3-by-5 card and held in front of you. In your mind's eye, see it in enough detail that you can discern each of the five symbols and their order in the equation.

2. Move the image around – left, right, up, down, towards you (in), away (out) – while keeping it whole and clear.

3. Zoom in and out. Make the image of the equation half its original size, and twice its original size.

If you can do the above, and feel comfortable with it, you can continue to the next section now. If you are having difficulty with it, you have some options. One is to continue forward – it may click into place, especially when you do the chapter exercises. Another option is to stop reading for a day. Tomorrow, practice the above three steps to refresh yourself, and to let you get more comfortable with it. When learning something new like this, giving yourself time to absorb it can be helpful.

2.2 Chunking

This section is about a new verb: *chunk*. The action of chunking is a key to taking an image you can make in your mind's eye, and using it to solve the equation it represents. You probably have some idea of what it is from chapter 1, where it was introduced in the example. It is a tool we use to help rearrange the equation in ways to our liking. Before we can fully define chunking and how we use it, we must discuss a few concepts.

The first concept is the *symbol*. Math equations are composed of symbols. In the equation $x + 2 = 3$, one symbol is the variable, 'x'. Two of the symbols represent mathematical processes ('$+$' and '$=$'), and the other two symbols are numbers. They are arranged in a particular order, giving the equation a particular meaning. If you change the order, it may change the meaning of the equation ($x + 3 = 2$) or it may not ($2 + x = 3$).

The next concept is an *object*. What exactly is an 'object'? How do you know something is an object, and how do you know that two objects are distinct? There are things in your environment right now that can be referred to as objects. You are wearing several[3], and if you are reading this on a computer screen, the display

---

[3]Well, you're *probably* wearing several. I don't know your reading habits.

is an object. Anything that you can pick up and move around is an object.

There are also non-physical objects. These are not real in the same sense as tangible objects, but they can still be useful. In particular, and this concerns us, math symbols are objects. Let's look at this equation again:

$$x + 2 = 3$$

Each of these five symbols can be considered an object. Whether a particular symbol is on a computer screen, paper, or in your mind's eye, you can consider them the same object.[4] In other words, the symbol 'x' on a computer screen is the same object as that 'x' on a printed page, and the same as an 'x' you imagine.

What about the equation itself? The equation as a whole is also an object. It is an object that is composed of other objects, the symbols. The same can be said for a jar of pickles, or a box holding kittens. The equation, jar, and box are objects that can hold other objects; when they are put in, the whole thing becomes a new, different, yet related object.

When you look at the equation above, you can focus on one symbol, an expression (group of symbols), or the equation as a whole. Your choice of focus will depend on what you need to do in that moment. When you choose to focus on an expression, or the entire equation, you are taking several objects and treating them as a single whole. This is fundamentally what *chunking* is. It is about working with an aggregate object that is composed of other objects. It implies a degree of flexibility in your thinking, since you are choosing to focus on an expression (such as $x + 2$) when it us useful to do so, on the equation as a whole ($x + 2 = 3$) when that is useful, or on one of the components individually (e.g., just x or 3) when THAT is useful.

---

[4]You can also consider them to be different objects. Neither view is more 'correct'. In this book, it is more useful to act as if a symbol like x is the same object regardless of where we see it.

Chunking is more about choosing where you put your attention than anything else.

The definition of *chunk* is "the action of consciously taking a group of objects, and working with that group as if it is a single object." We would say something like "Chunk the (some group of symbols)". For example, in the equation

$$\frac{9}{4+x} = 3$$

we can say "Chunk the $4 + x$", which means we are going to work with the expression $4 + x$ as a group. In this book, chunks are pointed out by drawing a dotted box around them.

$$\frac{9}{\boxed{4+x}} = 3$$

2.3 Exercises

Visualization

Pick three algebraic equations that you normally work with, or are similar to equations you typically use. You can choose them from a math textbook if you are a student, or from a research or work project if you are not. Pick equations that you consider to be on the simple side of medium complexity – that is, equations that are complex and 'big' enough to be interesting, but not much bigger.

It's also best to pick equations that are fairly different from each other.

When you've chosen them, do these exercises with each equation:

1. Visualize the equation as if it was written on a 3-by-5 card and held in front of you. In your mind's eye, see it in enough detail that you can discern each of the symbols and their positions in the equation.

2. Move the image around – left, right, up, down, towards you (in), away (out) – while keeping it whole and clear.

Chunking

For each of the following, visualize the equation with the indicated expression chunked. You may first write down the equation with the chunk outlined if it helps you.

1. $4x + 3 = x - 7$; chunk the x on the right side.

2. $\frac{2x^2+3x-4}{x+1} = 2$; chunk the $x + 1$ in the left side's denominator

3. $3 + \sqrt{x + 2} = 5.4$; chunk the 3 on the left side.

4. $3 + \sqrt{x + 2} = 5.4$; chunk the $\sqrt{x + 2}$ on the left side.

5. $3 + \sqrt{x + 2} = 5.4$; chunk the $x + 2$, inside the square root, on the left side.

Chapter 3

Symbol Motion

Let's take a simple equation, $2 + x = 3$, and examine the steps in solving it. It would go something like

$$2 + x = 3$$

$$2 + x = 3$$
$$-2 \quad\quad -2$$

$$x = 3 - 2$$

$$x = 1$$

Notice in particular the first step and the next-to-last step:

$$2 + x = 3$$

$$x = 3 - 2$$

There is a pattern here. The 2 is a symbol that is, in effect, moved from the left side of the equation onto the right side. There is a logic to it, which means there are rules to be followed – for one

thing, the 2 becomes a -2 as it is moved, in this case. It turns out
that there are patterns like this for many algebraic manipulations
you might use to solve an equation. By taking symbols, grouping them in particular ways, and moving them according to certain
rules, you can solve equations. This is called *symbol motion*. Doing algebra in this way is not well suited to working equations on
paper or a whiteboard. But it is very well suited for solving equations by visualizing them. This chapter explains the patterns of
symbol motion, applied to algebra, and how to use them.

3.1   Visual Logic

This expression

$$2 + x$$

and this expression

$$x + 2$$

are the same. You probably learned this by memorizing a rule,
called the *commutative property*, which states that the sum of several terms is the same regardless of their order. Because the rule is
so simple, most people instantly know that the two expressions are
equivalent, and can swap one for the other without much thought.

There is a way this process can be modelled visually. Start with
this expression:

$$\boxed{x + 2}$$

Visualize it, so that you see it mentally. In your image, chunk
the x and the 2 separately.

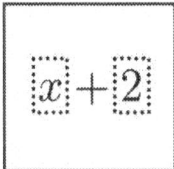

Now, rotate them around the + sign.

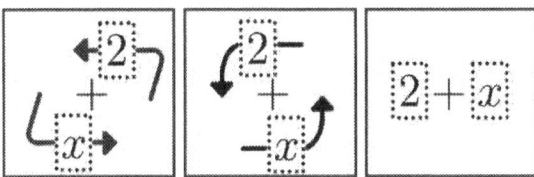

While we rotated it counter-clockwise around the +, there are other possibilities. You can rotate it the other way, or in the third dimension by having one symbol pass in front and the other pass behind. They can even 'teleport', immediately swapping places with each other. Try these out right now. Notice which one seems more natural for you.

The conversion above is a hint of something: we can take an image that has mathematical meaning, and by doing certain graphic manipulations on the image itself, end with the expression in a different, equivalent form. Done correctly, the graphic manipulations have the same effect as normal mathematical methods. Visual logic has to do with the abstract pattern of the change in the image. If the expression above was $y + 3$, we would transform it to $3 + y$, by shifting things around in a similar way to what we did above. Even though the symbols in that expression are different, the visual logic of the transformation would be the same.

When we do a transformation on an equation's image like this, there are two things going on at once. There is the visual logic, the pattern of change in the image itself. There is also the mathematical logic; $x + 2$ really does equal $2 + x$. A pattern of visual logic is really only useful to us if it completely honors the mathematical logic that is represented. As techniques are presented in this and the following chapters, keep in mind that this connection between the visual and mathematical logic has to be maintained in everything we do.

3.2   Movement As Arithmetic

If you solve this equation,

$$\frac{2x - 7}{3} = 5$$

you might multiply each side by three,

$$3 \times \left(\frac{2x - 7}{3} \right) = (5) \times 3$$

$$2x - 7 = 15$$

then add seven to each side,

$$(2x - 7) + 7 = (15) + 7$$

$$2x = 22$$

and finish by dividing each side by two:

$$\frac{(2x)}{2} = \frac{(22)}{2}$$

$$x = 11$$

These three arithmetic operations can be modelled in a particular, visual manner. When done correctly, they have the same mathematical effect as the "normal" way. Let's examine each operation in detail.

Categories of Symbols

Let's examine one of the intermediate equations above:

$$2x - 7 = 15$$

There are two kinds of symbols here.

1. *Quantities.* A quantity is a number or a variable that represents a number. 2, x, 7, and 15 are all quantities.

2. *Operations*, which are symbols that represent a mathematical operation. The only one in this expression is +; symbols like $-$, $\times$ (for multiplication), and $\div$ are also in this category.[1]

Operations have an interesting property: they need quantities in order to exist. We can have expressions like

$$4x + 9$$

but not expressions like

$$4x+$$

If the 9 is "taken away" because, say, we subtract 9 from both sides of an equation, the expression simply becomes $4x$.

Symbol motions can be classified as being in the *addition/subtraction* class or the *multiplication/division* class. The classes are named this way because of what their symbol motions do to the equation. Symbol motions in the addition/subtraction class have the same effect on an equation as if you add or subtract a term from each side; and those in the multiplication/division class have the same effect as if you multiplied or divided each side by something. Symbol motions within each category are very similar to each other. Let's look each class in detail.

---

[1] The equal sign would be in a third category, called *relations*. For now, we are just concerned with the first two categories.

Addition/Subtraction Class

In the second step, the equation goes from

$$2x - 7 = 15$$

to

$$2x = 15 + 7$$

This suggests a pattern of moving symbols, which has the same effect as the normal method of adding 7 to each side. The visual logic is

1. The -7 is moved from one side of the equation to the opposite side, *and*

2. its sign is flipped – that is, changed to its opposite. It goes from -7 to $+7$.

Let's see how you can apply this pattern, using symbol motion. Visualize this equation, like you practiced before:

$$2x - 7 = 15$$

Next, chunk the -7 – put your attention on the '-7' as a separate object, floating in the containing equation.

$$2x \boxed{- 7} = 15$$

Notice that we are taking two symbols here, the $-$ and the 7, and grouping them together in a single chunk. Next, move that chunk:

$$2x \underset{-7}{\overbrace{\quad = 15}}$$

While the -7 is in motion, flip (change) its sign:

$$2x \quad = 15$$
$$+7$$

When the motion completes, so does the arithmetic operation:

$$2x = 15 + 7$$

Here is a summary of what we just did:

1. Visualize the equation.

2. Chunk the term (symbol or symbols) you wish to move, including the $+$ or $-$ in front of it. (If there is no $+$ or $-$ there, insert a $+$ in front of it.)

3. In the image you are visualizing, begin to move the chunk to the opposite side. As you do this, flip the sign inside of the chunk. If it has a $+$ in front, replace that with a $-$, and vice versa.

4. Move the chunk into the opposite side of the equation. Tack it onto the far right. (You can also add it in between other terms that are being added or subtracted. If you put it on the far left, place a $+$ sign between it and the next term.)

You can verify that you moved the symbols correctly, by doing the equivalent operation the "normal" way – by adding or subtracting an expression on each side – and check whether the answer is the same.

Multiplication/Division Class

The visual logic for the multiplication/division class is simpler than that for the addition/ subtraction class. Again, we will show how it is done with an example. In the steps at the beginning of this chapter, we go from

$$\frac{2x+7}{3} = 5$$

in the first step, to

$$2x + 7 = 5 \times 3$$

in the second step. First, visualize the equation:

$$\boxed{\quad \frac{2x+7}{3} = 5 \quad}$$

Then, chunk the 3 like this:

$$\boxed{\quad \frac{2x+7}{\boxed{3}} = 5 \quad}$$

Start to move the chunk to the other side.

$$\frac{2x+7}{} = 5$$

As you can see, the first steps are the same: visualize the equation, chunk a symbol or expression, and start to move the chunk to the opposite side of the equation. The next steps are slightly different. When we move chunks in this way – meaning, in a way that is equivalent to multiplying both sides of the equation by that symbol – we are going to move it from the *denominator* or the bottom of one side, to the *numerator* or top of the other side. (By the way, it works the other way too. If we are moving a chunk that starts in the numerator of one side, we would move it to the denominator of the other side.)

What happens when the other side is not a fraction? Actually, it is. In this example, the right hand side is just 5. It is also the fraction $\frac{5}{1}$. Keep in mind that every expression is also a fraction in this way.

As we move the 3, we do two special things. First, we remove the horizontal line that used to be between "$2x + 7$" and "3":

$$2x+7 = 5$$

In this image, the horizontal line is a symbol. In particular, it is an *operation*, in the sense described on page 25. Remember that an operation needs two quantities in order to exist. When we take one of those quantities away, the operation symbol vanishes. That is what has happened here.

Next, we insert a $\times$ symbol (that's "times", for multiplication, and not the letter x) to the right of the five:

$$2x + 7 = 5 \times$$

Finally, we move the 3 into place:

$$2x + 7 = 5 \times \boxed{3}$$

There is another way to do all this. After removing the horizontal line,

$$2x{+}7 = 5$$

we can move the 5 to the right, and put the $\times$ (multiplication sign) to its left:

$$2x + 7 = \ \times 5$$

Then we move the 3 into the space to the left of the $\times$:

$$2x + 7 = \boxed{3} \times 5$$

Both "pathways" are fine. They are both equivalent to multiplying each side by 3, and that is all that is important.

Division is similar to multiplication. Let's go backwards, from

$$2x + 7 = 5 \times 3$$

to

$$\frac{2x + 7}{3} = 5$$

With multiplication, we move the symbol from the denominator on one side to the numerator on the other side. With division, it is the opposite. We are taking a term from the numerator (or just from the expression, if it is not a fraction) on one side, and moving it into the denominator on the other side.

Start by chunking the 3.

$$2x + 7 = 5 \times \boxed{3}$$

Begin to move the 3 across.

$$2x + 7 = 5 \times$$
$$\boxed{3}$$

As it moves, we do two things. First, we drop the multiplication symbol (the $\times$) on the right side.

$$2x + 7 = 5$$
$$\boxed{3}$$

Also, since we are going to make the left side into a fraction, we need to bring in the operator. In other words, we need to draw a horizontal line under the numerator:

$$\frac{2x+7}{\boxed{3}} = 5$$

(Note that in this situation, "bring in the operator" and "draw the horizontal line" are two ways of saying the same thing.) This done, we finish moving the 3 into the denominator of the left side.

$$\frac{2x+7}{\boxed{3}} = 5$$

Combining arithmetic types

Here is a more complex example, which is solved with a combination of the above methods. First we will briefly solve it using the normal algebraic methods. Then we'll solve it again using symbol motion. In this equation,

$$\frac{3x + 7}{2x + 1} = \frac{5}{2}$$

one way to solve it is to first multiply each side by 2:

$$2 \times \left(\frac{3x + 7}{2x + 1}\right) = 2 \times \left(\frac{5}{2}\right)$$

$$\frac{2 \times (3x + 7)}{2x + 1} = 5$$

Then, multiply each side by $(2x + 1)$.

$$(2x + 1) \times \left(\frac{2 \times (3x + 7)}{2x + 1} \right) = (2x + 1) \times 5$$

$$2 \times (3x + 7) = (2x + 1) \times 5$$

$$6x + 14 = 10x + 5$$

Collect the variables on one side and the numbers on the other, and simplify.

$$6x + 14 - 6x = 10x + 5 - 6x$$

$$14 = 4x + 5$$

$$14 - 5 = 4x + 5 - 5$$

$$9 = 4x$$

$$x = \frac{9}{4}$$

Now, how do we do the same thing through symbol motion? Before we start, it's good to point out a few things. First, when moving symbols, we obey the normal rules of arithmetic. For example, if we are solving this equation:

$$\frac{2}{x + 3} = 1$$

we would chunk the $x + 3$

$$\frac{2}{\boxed{x+3}} = 1$$

and begin to move it to the right side.

$$\frac{2}{\boxed{x+3}\rightarrow} = 1$$

We would certainly NOT do something like this:

NO!

$$\frac{2}{+3} = 1 \qquad \boxed{x}\rightarrow$$

We can always be clear by asking what sequence of arithmetic operations that the symbol motion is supposed to emulate. Here is a general principle:

Every symbol motion has the same effect as a sequence of mathematical operations.

If (and only if) some change in the image is equivalent to some such sequence, is that change a valid symbol motion. If there is no such sequence of operations, the change is not a symbol motion and has no mathematical meaning.

Let's work the equation at the beginning of this section again, this time using symbol motion.

$$\frac{3x+7}{2x+1} = \frac{5}{2}$$

We'll start by making two chunks, both of which will be moved:

$$\frac{3x+7}{2x+1} = \frac{5}{2}$$

Both of these chunks are in the denominators. We move each of them to the numerator of the opposite side:

$$\frac{3x+7}{2x+1} = \frac{5}{2}$$

As we do this, we erase the horizontal lines, and insert multiplication symbols ($\times$) as appropriate.

$$2 \times (3x + 7) = 5 \times (2x + 1)$$

Notice that we have also put parentheses around two of the expressions. This is sometimes needed when symbols are moved around, for the same reason they are needed in normal algebraic manipulation. Next, we distribute the multiplication on each side:

$$6x + 14 = 10x + 5$$

Chunk the $6x$ and the 5, and begin to move them.

As we do this, we switch the signs of each, from $6x$ to $-6x$ and 5 to -5.

Normal arithmetic quickly solves the rest ($x = \frac{9}{4}$).

3.3 Summary: The Rules of Symbol Motion

There are two classes of symbol motions: those that are equivalent to an addition or subtraction operation, and those that are equivalent to a multiplication or division operation.

In the addition/subtraction class, you start by chunking the expression you want to move. If there is a + or a − to the immediate left of the expression, you include that symbol in the chunk. As you move the chunk, you multiply its contents by −1 (i.e., just flip its sign). You normally insert the contents of the chunk at the far right of the opposite side. You can also insert it in another location, if you add it in between terms that are being added or subtracted together.

For the multiplication/division class, start by chunking the expression you wish to move. If it is in the numerator of its current side, you will move it to the denominator of the other side; if it is on the denominator, you move it to the opposite side's numerator. (If one side does not look like a fraction, it actually is – just put it over 1.)

3.4 Exercises

Do the following with each equation below:

1. On paper, solve the equation for x using the normal methods, writing down the intermediate steps.

2. Next, also on paper, solve the equation using symbol motion. Outline the chunks, and use arrows to point out what is moving and where. Write (draw) each step in which a symbol is moved.

3. Solve the equation using symbol motion, with your visual imagination – in other words, without writing it down. If you

at first need to write down some of the intermediate steps, or use what you wrote in steps 1 and 2 as a guide, you can do so. Just make sure to use that to support your learning, not to "cheat". Repeat the process until you are able to solve the equation mentally without those aids.

Here are the equations:

1. $2x + 3 = 7$

2. $\frac{3x-5}{2} = 5$

3. $x + \frac{1}{2} = 2x - 4$

4. $\frac{3}{x-1} = 2$

5. $2 = \frac{2x-4}{1-x}$

Chapter 4

Some Advanced Techniques

Having been introduced to the basics, there are two techniques you can learn now. The first is called *windowing*, and has been mentioned briefly in chapter 1. It is a tool that helps you work with images faster, and work with equations that are larger than you can easily visualize. The second technique is called *correspondence*. You will use this simple concept and skill when solving quadratic equations, using expressions with exponents and logs, and other situations.

4.1   Windowing

You now have some experience visualizing equations, and working with those images. This is important. As you can tell by now, pretty much everything you learn to do in this book starts with seeing math symbols in your mind's eye.

If you needed to envision the whole equation completely in order to solve it, you could sometimes run into trouble. What if you want to solve an equation that is just too big for you to visualize at this point? Even when you have more practice and become very good at visualizing – or if you are naturally talented at it – bigger and more complex equations will always exist.

Fortunately, you can sidestep all this using a technique called *windowing.* To use windowing, you do need to be able to remember the equation. The rule of thumb is that you will need to remember it well enough that you could write it down (though you normally would not). At the same time, you would only have to be able to visualize a portion of it at once. Note we are making a distinction between two mental capacities you have: your memory, and your ability to visualize. Most people will find they can exactly remember equations that are much larger than the biggest one they can visualize. Using this ability, you can mentally see enough of the equation to begin processing it.

For equations you can visualize easily enough, you may still want to use windowing, because doing so will let you solve the equation faster and more easily. I actually use windowing for almost every equation I solve mentally, except for very small ones.

What portion of the image (equation) do you need to be able to visualize? It depends. As you have learned, you solve the equations by manipulating the image in particular ways. Normally you are only changing part of the image at any one time. You apply windowing by visualizing those parts of the equation that need to be manipulated, applying the operations to them, and then translating the effect back into your memory of the equation. The key is to become flexible at calling up an image of those parts when you need them. For example, pretend you are solving this equation.

$$\frac{4x + 2 + \sin 2\pi x}{3} = 1 + 2x + \frac{x^2}{2}$$

Perhaps you can visualize this whole equation, perhaps not. Let's see how the use of windowing can help us solve this equation without visualizing every bit of it. The next obvious step in solving it is to multiply each side by 3. The way to do this is to visualize the equation, and chunk the 3 on the left side's denominator.

$$\frac{4x+2+\sin 2\pi x}{3} = 1 + 2x + \frac{x^2}{2}$$

Then, you would move the chunk to the other side like this:

$$\frac{4x+2+\sin 2\pi x}{3} = 1 + 2x + \frac{x^2}{2}$$

$$4x + 2 + \sin 2\pi x = 3 \times \left(1 + 2x + \frac{x^2}{2}\right)$$

The thing to notice is that only part of the image (the 3 in the left side's denominator) moved. Most of the image did not change at all. Whenever this happens, we have a ripe opportunity to use windowing. See if you can visualize this:

Those are just blurs where the $4x + 2 + \sin 2\pi x$ and $1 + 2x + \frac{x^2}{2} + x^4$ used to be. A lot easier to visualize, isn't it? The essence of windowing is to choose to visualize portions of the equation – in this case, the chunked 3 and a few symbols that show the equation's structure – and purposefully allowing the rest of the equation to be foggy. You can visualize those fogged-out parts later, if and when you need them.

You don't have to imagine blurry bars there, by the way – you can visualize squiggly purple lines, empty space, or something else. Experiment, and do what works best for you.

You can move the chunk just like we did a few images ago:

You then "bring in" the other symbols as needed. That is, when you are ready to use them, you can take parts of the equation that are foggy and fully visualize them. Since a good next step is to distribute the multiplication on the right hand side, we can bring the symbols that are in the parentheses in next.

$$\rule{3cm}{0.3cm} = \boxed{3} \times (1 + 2x + \tfrac{x^2}{2})$$

Then you can de-chunk the 3 and multiply out the right side like normal.

$$\rule{3cm}{0.3cm} = 3 + 6x + \frac{3x^2}{2}$$

Alternatively, you can un-blur each of the three terms – 1, $2x$, and $\frac{x^2}{2}$ – one at a time, multiplying each term by 3 before getting the next.

Windowing is very natural once you get used to it. In fact, you may already have started doing it before reading this chapter, perhaps without realizing it. If you are not at that point yet after working this chapter's exercises, you can get there by introducing it gradually. When you solve equations using symbol motion, visualize the whole equation at first if you need to. Notice if there are parts of the equation you are able to **not** visualize, even briefly. You can bring those parts back "in focus" when they are needed to further solve the equation.

4.2 Correspondence

What you have learned so far has all involved taking an equation's image and making incremental changes or manipulations to it. There are instances, however, in which you will need to make some wholesale change to the entire equation. A good example is in solving certain equations containing logarithms, such as this one.

$$3\log(x + 1) = 6$$

(This is a base 10 logarithm.) Using symbol motion would get you here:

$$\log(x + 1) = 2$$

To go further, you would need to use the property that states that $\log_b a = c$ means $b^c = a$. *Correspondence* is when you construct a new image, based on information such as this, that contains the equation in a different form. In this case, you would first have an image like this,

$$\log(x + 1) = 2$$

and substitute it with an image like this,

$$10^2 = x + 1$$

which you can use normal arithmetic and symbol motion to solve. Another example of this process is when you use the quadratic formula to solve an equation.

$$2x^2 - 3x - 9$$

You would substitute that image with this one.

$$\frac{3 \pm \sqrt{(-3)^2 - 4(2)(-9)}}{2(2)}$$

This is nothing new – everyone learns to do things like this in their first or second algebra course. What is new is that you are doing this kind of equation-wide substitution mentally. You are applying it to the representation of the equation you have in your mind's eye. It may take practice before you can do it well (though it's also possible it will happen for you immediately). It is a simple process, and by the time you have some experience at doing this, it will be second nature.

4.3 Exercises

Windowing

Choose three equations that are big enough that you can remember (meaning you can write them down by memory) but are bigger than you can easily visualize. (If you can visualize all equations you can remember, just choose three of the largest equations you can visualize.) You will get the most out of this exercise if you pick equations that are fairly different from each other. Write down these equations so that you can refer to them. You will also need another blank sheet to write on.

Do the following with each equation.

1. Look at the equation, and study it so that you can remember it. Turn over the sheet or cover it, so you can't see it. On your other sheet, write down that equation from memory. When

you are done, check that you remembered it 100% correctly – that's important! If what you wrote down does not match, or you are not able to remember it, study it again and repeat until you are able to do this.

2. Where you wrote down the equation, divide it up into three to five expressions, circling each one. You want to choose expressions that you are able to visualize (since that is what you will be doing with them). When you are done, every quantity in the equation should be in a circled group. For example, if your equation is $\frac{x^2+3x+1}{x-1} = 4 + 2\sin\frac{\pi x}{4}$, you might circle the $x - 1$ on the bottom left, the $x^2 + 3x + 1$ on the top left, the 4 on the right, and the $2\sin\frac{\pi x}{4}$, for a total of four circled expressions. (You don't necessarily need to circle all the operators.)

3. Cover the equation so that you can't see it (or close your eyes, etc.) Visualize the structure of the equation, like in the figures on pages 41 to 43, so that you are able to see its general outline (even if you are not visualizing each and every symbol involved). Include in the image one of the expressions you circled before. You will end up visualizing something that has the general structure of the full equation, with the circled expression visualized and the remaining symbols blurry or just not visualized. Practice if you need to, until you can mentally see this. You can write/draw a diagram (like the figures mentioned above) to help you if needed.

4. Next, allow that circled expression to fade out. Pick one of the other circled expressions, and bring it into the image instead. You are doing something exactly like in the previous step, just with a different circled expression. Practice if needed, until you are able to mentally see this. Repeat this process with each remaining circled expression.

Correspondence

1. Take each of these equations, and using symbol motion, reduce them to the form $\log_b a = c$. Then visualize them in the converted form, $b^a = c$, and finish solving the equation from there. (For example, take the equation $2 \log x = 4$. First you simplify it to $\log x = 2$, using symbol motion. Then you visualize it in its converted form, $10^2 = x$, or $x = 100$.

   (a) $3 \log(x + 1) = 3$

   (b) $-2 + \ln x = 4$

   (c) $3 + 2 \log(2x + 20) = 7$

2. Find the root of each quadratic equation below. (Since we haven't covered how to deal with multiple solutions yet, just find the more positive root, $x = \frac{1}{2a}\left(-b + \sqrt{b^2 - 4ac}\right)$.) Solve it by rearranging the equation to the form $ax^2 + bx + c = 0$, using that to construct the equation for the more positive root above, then solving from there. Do all of this mentally, using symbol motion, correspondence, etc. For example, if the equation is $2x^2 - 3 = x$, you will visualize it, use symbol motion to change it to $2x^2 - x - 3 = 0$, then convert that to

$$x = \frac{1}{2(2)}\left(-(-1) + \sqrt{(-1)^2 - 4(2)(-3)}\right) = \frac{3}{2}$$

   Since visualizing that whole expression may be tricky, you can visualize and solve just $\sqrt{b^2 - 4ac}$ first. Then visualizing the rest is easier: $x = \frac{1}{2(2)}(-(-1) + 5) = \frac{3}{2}$.

   (a) $x^2 + 2x - 3 = 0$

   (b) $-x^2 - 4x = 3$

   (c) $6x - 5 = x^2$

   (d) $-2x^2 - 8x = 6$

Solutions: 1(a) $x = 9$, (b) $x = e^6$, (c) $x = 40$. 2(a) $x = 1$, (b) $x = -3$, (c) $x = 1$, (d) $x = -3$

Chapter 5

Morphing Symbols and Expressions

When solving an equation, we often do some operation that involves rearranging the symbols with minor alterations. Symbol motion covers this kind of operation. Other times, we need to replace a few symbols with something else entirely. This happens when we are simplifying or reducing part of the equation. Some simple examples are if we take an expression $4x - x$ and replace it with $3x$, or if we take $\sqrt{x^2}$ and simplify it to x. An alteration like this, applied to the image of an expression, is called a *morph*. A morph is when we take a group of symbols in the image of an equation, and replace it with a smaller or simpler group of symbols that is mathematically equal. The concept is not a big deal; that's exactly what you are doing when you simplify or reduce something, and you do that all the time. So why do we have a whole chapter about it? The difference is that while substitution is a mathematical action, morphing is a visual action. Morphs are often used when doing some kind of substitution, in a broad sense of the word. There are some patterns we can use that make the task easier. This chapter explains them and how to use them. We start with a kind of morph called *fusion*, and then introduce more general morphs.

5.1   Fusion

Something you will do a lot is take a group of terms and reduce it, replacing the original expression with a smaller one. The simplest example of this is basically arithmetic: taking '$2x - x$' and replacing it with 'x', for example, or replacing '$1.3 + 2.2$' with '3.5'. The act of combining symbols like this is called *fusion* (or *symbol fusion*, if you want to be verbose.) A fusion is probably the most common kind of morph you'll use. The name is an analogy to nuclear fusion, which happens when two atoms combine to form a single, different atom. When you visualize the equation, you focus on that part of the image that represents the expression you want to change. You visualize the symbols actually moving together, and shifting form into the reduced expression.

As said above, a fusion takes place when you add (or subtract) several terms together to make one term. Another kind of fusion takes place here:

$$\frac{3x + 2 - 2}{x} = \frac{3x}{x} = 3$$

This has two fusions, one of which is when the $3x + 2 - 2$ is simplified into $3x$. The other fusion is a new type, where the x's cancel out in $\frac{3x}{x}$. It may not be obvious that the second one is a fusion. Anytime we take a group of symbols, and replace them with a mathematically equivalent group of fewer symbols, we have fusion. If the expression happens to reduce to zero (when it's being added) or one (when it's being multiplied), so that it would be replaced with nothing, we still call that fusion. This means cancelling out terms in fractions is a kind of fusion.

When you visualize the equation, you can actually visualize the symbols fusing together and morphing into the replacement expression. For an expression like $2y + 4 - y$, you start by visualizing it,

$$2y + 4 - y$$

then use symbol motion to collect the y variable terms,

$$2y + 4 - y$$

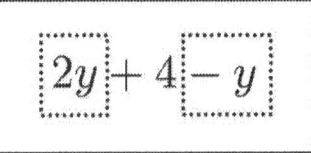

$$2y - y + 4$$

$$2y - y + 4$$

then chunking them.

$$2y - y + 4$$

Next you imagine the symbols combining:

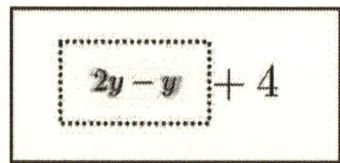

You then visualize the fused symbols morphing into the re-placement:

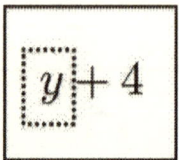

By the way, when we show an image like this,

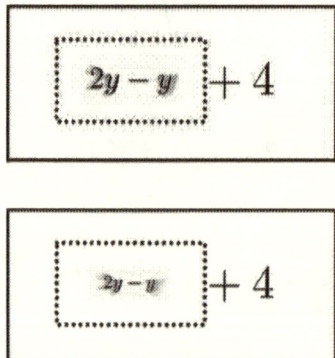

it is really shorthand for a sequence like this:

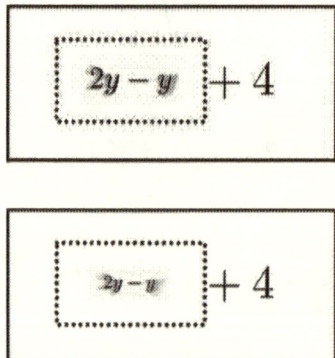

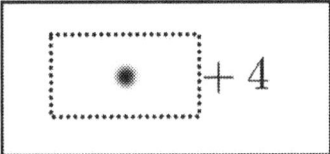

Keep this in mind when you see it later. Imagine the expression collapsing in on itself.

Let's note a few things. First, fusion will not do the math for you. You need to be able to calculate that $2y - y = y$. What fusion does is help you focus on changing that particular part of the image (and the equation).

Second, the details of how this is done are somewhat flexible. You don't necessarily have to visualize the process as shown in these figures. The only requirements are that you (a) start with seeing the expression in it starting form, and (b) end with seeing the expression in its reduced (fused) form. So long as you honor those criteria, you can see them moving together and transforming in any kind of sequence or pattern. It may or may not resemble what is shown in this chapter. Maybe you will see the symbols slurping together and reforming in an animation. Or maybe you will not want to have an animation at all; you will instantly substitute the reduced form, or have it go through one or a few intermediate images. Try a few different methods and decide which works best for you.

For the expression $\frac{3x+2-2}{x}$, you start by visualizing it,

$$\frac{3x+2-2}{x}$$

then fusing the numerator,

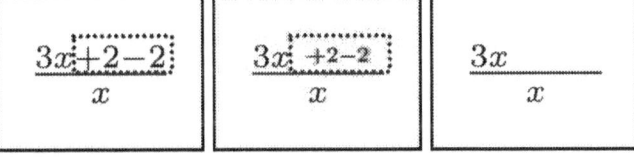

The next step is to cancel out the x. You can envision this as the $\frac{x}{x}$ part caving in on itself, morphing into nothing.

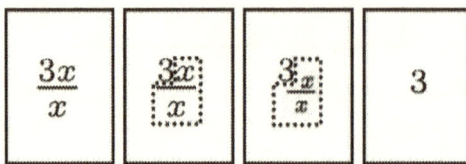

See how the chunk has an irregular shape there. You can do that.

Instead of visualizing the fusion happening as shown above, you can also see the parts being struck out, then disappearing:

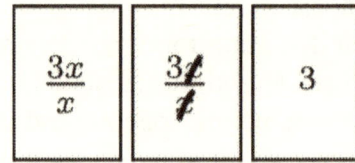

As is often the case when you are cancelling a term, it is simple enough that you can just do it mentally without any special technique. Thus, you can simply allow the $\frac{x}{x}$ drop or fade from the image. The two methods shown here are basically ways of doing that same thing.

Fusion takes place at different levels. The fundamental idea is to take a group of symbols and *combine* them into a smaller, simpler group. (Of course, they have to be mathematically equal.) The examples of fusion above are basically arithmetic. While this simple level is important, fusion gets more interesting as you use it to make larger 'jumps'. In the example above, if we break it down and write every dinky little step we can think of, it might look like

$$\frac{3x + 2 - 2}{x} = \frac{3x + 0}{x} = \frac{3x}{x} = 3 \times \frac{x}{x} = 3 \times 1 = 3$$

We skipped most of these steps, and when you process it internally, you likely skip them too. Your level of mastery of math is such that, for example, you can go directly from $3x + 2 - 2$ to $3x$. You probably did not stop to think of $3x + 0$ explicitly. As an analogy, pretend you are crossing a shallow pond, by stepping on a series of stones. Many of the stones are close together, and you can step on each one if you want. If you have long legs and good balance, you can step on every other stone, or even jump across three or eight at a time, and make it safely across.

You can stretch and pace yourself. Some days, you will be able to take huge jumps – doing several steps at once, except that it will *seem* like just one step in your mind. The first time you notice this, congratulate yourself, because that is exactly how a talented mathematician does things.

Conversely, if you are tired, you may have trouble taking the kind of jumps you are used to. If that happens, you can just take smaller steps, so to speak. Instead of going from $\frac{3x+2-2}{x}$ directly to 3, take some or all of the intermediate steps above as you work the equation's image.

There is also symbol *fission*, which is the opposite of symbol fusion. If we replace $2x$ with $4x - 2x$ or even $5x - 4x + x$, that is fission. You won't split up symbols like this nearly as often, but sometimes doing so is useful, such as when you are factoring polynomials.

5.2   Other Morphs

Using symbol motion and fusion will let you solve many equations intuitively. There are also situations where you need something else. Take this equation:

$$(x - 2)^2 = 4$$

The easiest solution path involves taking the square root of each side[1]:

---

[1]More properly, we'd end up with $x - 2 = \pm 2$. We'll just work with the

$$\sqrt{(x-2)^2} = \sqrt{4}$$

$$x - 2 = 2$$

There is not a way to express square roots using symbol motion, fusion or fission. We need another kind of morph. We are now getting into an area where the technique is less structured than what you've seen so far. The benefit is that such morphs are more flexible, and even somewhat adaptable.

Let's make this concrete by showing how morphs are used to solve square roots. The process goes something like this:

1. Set up the equation so that you can apply the square root to both sides.

2. Add symbols to each side that describe that operation.

3. Simplify each side, one at a time, using all the techniques at your disposal – symbol motion, fusion, or other morphs.

Here is an example of how that works. As always, first visualize the equation.

$$(x-2)^2 = 4$$

The first step in applying the morph is to visualize a square root around each side:

---

positive root here, since we haven't learned how to deal with equations with multiple solutions yet. (That comes in chapter 6, "Plurality".)

$$\sqrt{(x-2)^2} = \sqrt{4}$$

Next, in the image, we substitute the changed expression, on one side at a time. So starting with the image just above, morph on the left side first,

$$\sqrt{(x-2)^2} = \sqrt{4}$$

$$\sqrt{(x-2)^2} = \sqrt{4}$$

$$x - 2 = \sqrt{4}$$

Then we do the other side.

$$x - 2 = \sqrt{4}$$

$$x - 2 = \boxed{\sqrt{4}}$$

$$x - 2 = 2$$

Instead of a graphical substitution, you can also do an animation. So with the change from $\sqrt{(x-2)^2}$ to $x-2$, you can imagine the $\sqrt{(\ \ )^2}$ part fading out. The animation from $\sqrt{4}$ to 2 will be more fanciful; most morphs will be like that.

Another use of morphs is when working with equations that use exponents, and which you would solve using logarithms. The process is quite similar to what we just did with the square root. For example, say you have this equation,

$$A(t) = A_0(1 + e^{a+bt})$$

Here A_0, a, and b are constants, t is the independent variable, and $A = A(t)$ is the dependent variable. This one is very similar to equations like $A(t) = A_0 e^{bt}$, which you have probably seen at some point in a textbook (or will see). You want to invert this equation, so that instead of $A(t) = something$ you have $t(A) = something$. If it was the textbook form, you may actually remember the inverted form, but since it is different you will have to invert it yourself. Let's do that.

We start by visualizing the equation, as always.

$$A = A_0(1 + e^{a+bt})$$

Remember that you can use windowing to make visualizing it easier (page 39). Next, using symbol motion, rearrange the equation so that the exponential is alone on one side:

$$e^{a+bt} = \frac{A}{A_0} - 1$$

Now we will use the morph. Apply the natural log to each side.

$$\ln(e^{a+bt}) = \ln(\frac{A}{A_0} - 1)$$

Remember that there is a property of logarithms, that $\log a^b = b \log a$. Because of this, you can use symbol motion to bring the $a + bt$ in the exponent outside of the log:

$$\ln(e^{a+bt}) = \ln(\frac{A}{A_0} - 1)$$

$$a + bt \quad \ln e \qquad = \ln(\frac{A}{A_0} - 1)$$

$$(a + bt) \ln e \quad = \ln\left(\frac{A}{A_0} - 1\right)$$

(This is a use of symbol motion that you did not see before. You will occasionally find interesting new places you can apply it like this.) The next step is to deal with the $\ln e$, which you do by chunking it,

$$(a + bt)\,\boxed{\ln e} \quad = \ln\left(\frac{A}{A_0} - 1\right)$$

then morphing into nothing.

$$(a + bt)\,\boxed{\ln e} \quad = \ln\left(\frac{A}{A_0} - 1\right)$$

$$(a + bt) = \ln\left(\frac{A}{A_0} - 1\right)$$

From there, you can use symbol motion to complete the inversion:

$$t = \tfrac{1}{b}\left(-a + \ln\left(\tfrac{A}{A_0} - 1\right)\right)$$

Remember what was said in the last section, about crossing the pond using stepping stones? And that with experience, you will be able to take larger and larger 'jumps' in your algebra calculations? This scenario is one place where this can happen. Above, our process changed $\ln e^{a+bt}$ into $(a + bt)\ln e$, then to $a + bt$. After you have done this a while, you will go from $\ln e^{a+bt}$ to $a + bt$ in one step inside your mind. You will chunk the $\ln e^{a+bt}$, and morph that chunk directly into $a + bt$. This kind of progression can apply to all morphs (actually, any mental math you do at all). As you get more experience, you will discover you can do bigger substitutions and changes at once by using bigger or different morphs.

Morphs can be arbitrarily general. In other words, there are other kinds of morphs than what are shown in this chapter. Most of the morphs used in doing algebra work are introduced in this book. Other morphs and similar techniques can be used in other areas of math, such as calculus.[2]

5.3 Exercises

1. Visualize each of the expressions below. In your image, reduce the expression via fusion.

    (a) $x + 3 - x$

    (b) $y^2 + 4x - 2x$

    (c) $\frac{3x-1}{x} + \frac{1}{x}$

---

[2] The book *Inner Math: How To Do Algebra, Calculus, and more In Your Head*, which is a sequel to this one, will show how to do this.

(d) $x + y^2 + x - 2x$

2. Solve each of the following equations mentally. If you need to take a root, or do something else that would lead to multiple solutions, just pick the more positive value at that point.

   (a) $(x + 1)^2 = 16$

   (b) $x^2 - 10x + 20 = 11$ (Hint: Complete the square.)

   (c) $10^{2x-3} = 1000$

   (d) Invert this equation: $x(t) = \alpha + \beta e^{-ct}$

Solutions: 2(a) $x = 3$, (b) $x = 9$, (c) $x = 3$, (d) $t(x) = -\frac{1}{c} \ln\left(\frac{x-\alpha}{\beta}\right)$

Chapter 6

Plurality

Everything up to now has been focused on what we call *singular* equations. Singular equations are simple in a particular way: they are solved in a sequence of steps that unambiguously leads to a single solution. One equation that is NOT singular is $(x-1)^2 = 1$. It has two solutions, $x = 2$ and $x = 0$. When solving it, we morph the each side by taking its square root:

$$\sqrt{(x-1)^2} = \sqrt{1}$$

In doing this we get two possible values for the right hand side:

$$x - 1 = -1$$

$$x - 1 = 1$$

The techniques we've introduced so far all involve working with a single picture; having a 'split' like this adds some complexity. Sometimes we can adjust the situation in ways that let us continue with a single picture (for example, if we work with the symbol ± 1). Yet there are some cases, like what we have here, where that strategy will not get you all the way to the final solutions. Other methods – conveniently covered in this chapter! – can be employed to reveal the solutions in these situations.

The definitions for singular and plural equations given before have been a little vague. This is because the concepts needed to give a precise definition had not been fully introduced. Now we can say that an equation is singular if can be solved by starting with a single equation image, and by applying symbol motions, fusions/fissions, and other morphs, arrive at a single solution. The key word is *single*. If we get a situation like the above, where we have two or more valid choices for where the equation is headed, it's a plural equation.

As we work the equation above, we arrive at a 'split': a point where a normal, single image cannot represent the two or more eventual solutions. As stated, at that point, the equations are

$$x - 1 = -1$$

and

$$x - 1 = 1$$

This chapter describes two strategies for solving plural equations internally. We'll introduce the first one by using it to finish solving this equation.

6.1 Multi-images

The first method is to have what is called a *multi-image*, which is just a single image that has both equations after the split:

$$x - 1 = 1$$
$$x - 1 = -1$$

In other words, instead of just visualizing one equation, you visualize both of them. You then do the manipulations for each individually.

$$x \quad \boxed{-1} \\ = 1$$
$$x - 1 = -1$$

$$x \quad = 1 \boxed{+1}$$
$$x - 1 = -1$$

$$x \quad = 1 + 1$$
$$x \quad = -1 \\ \boxed{-1}$$

$$x \quad = 1 + 1$$
$$x \quad = -1 \boxed{+ 1}$$

Here we did each equation one at a time. If you are good, you can solve both simultaneously.

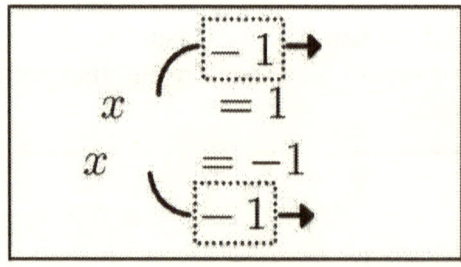

It is powerful to combine the use of multi-images with window-ing. Visualizing several variations of the same equation at once is a lot of work. Because of the similarity, though, it is relatively easy to window them.

6.2   Subimages

Multi-images can take a lot of mental energy, especially if your equations are complex or if there is more than one split. What can help is to take the concept of a multi-image, and apply it to a *portion* of the equation's image, in a way that keeps the logic intact. This often makes solving the equation much easier.

We'll illustrate by showing how to solve quadratic equations using the general solution. As an example, we'll use $2x^2 + 5x - 3$, and plug in the formula $x = \frac{-b \pm \sqrt{b^2 - 4ac}}{2a}$. The figure we visualize becomes

$$x = \frac{-(5) \pm \sqrt{(5)^2 - 4(2)(-3)}}{2(2)}$$

which reduces to

$$x = \frac{-5 \pm 7}{4}$$

The numerator has two legitimate values, $-5 - 7$ and $-5 + 7$. The basic idea of subimages is to have a single image that represents both solutions, as compactly as possible, while letting us solve for both. The dual image introduced near the start of this chapter partially satisfies all this.

$$x = \frac{-5 + 7}{4}$$

$$x = \frac{-5 - 7}{4}$$

Yet this image does not represent the information as compactly as possible. The most compact image would be what we wrote above:

$$x = \frac{-5 \pm 7}{4}$$

That's nice and all, but this image does not meet the third criteria – letting us continue solving for both values. (That is, at this point in solving the equation. Notice that prior to this step, it did.) What helps here is a new method of presenting the situation visually. Visualize just one of the solutions:

$$x = \frac{\boxed{-5+7}}{4}$$

I have taken the liberty of chunking the $-5+7$ for you. Notice that the outline is a little different. Pretend that the $-5+7$ is like a card that you can pick up and remove from the image:

$$x = \frac{\quad\quad}{4} \quad\longrightarrow\; \boxed{-5+7}$$

And you can drop another card in its place, representing the other solution:

$$x = \frac{\boxed{-5-7}}{4}$$

Now you have a modular system. Your image can represent either solution depending on which card you place on (in) it.

You can work with the image in the card just like any expression, using everything in this book.

$$x = \frac{\boxed{-5+7}}{4}$$

becomes

$$x = \frac{\boxed{2}}{4}$$

We call a card like this a *subimage*. It is a group of symbols in the equation that you mark off as separate. It is like a chunk, but more definite in its boundary, and more independent from the expression it lives in.

You can solve plural equations by using multiple subimages (cards). There are several ways to visualize this. One is to see the subimages partly stacked:

$$x = \frac{\boxed{\begin{array}{c}-5-7\\ -5+7\end{array}}}{4}$$

Here they are offset so that you can tell what is on both. You can also stack them directly on top of each other:

$$x = \frac{\boxed{-5-7}}{4}$$

These methods are both fine, and there are other ways too. Use the variation that works best for you.

When you want to work with the other solution, take that and swap it into the foreground of the image.

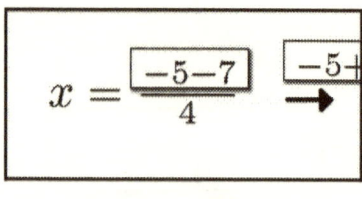

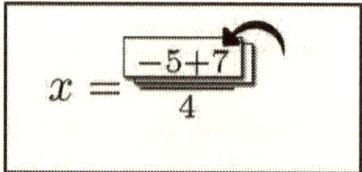

Having defined the subimages, you are set up to find all the solutions. It's usually best to work (simplify) within the subimages first, and when they are all done, move forward from there. We'll solve within the foreground subimage first.

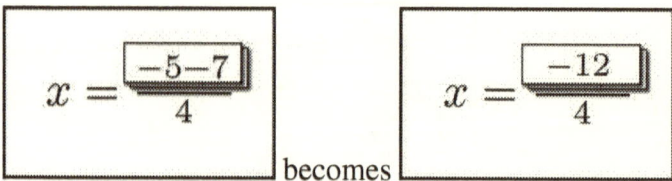

becomes

Now that the top subimage is done, bring the bottom one up and work it.

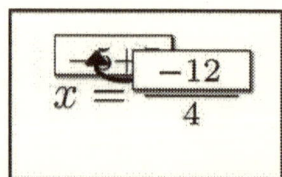

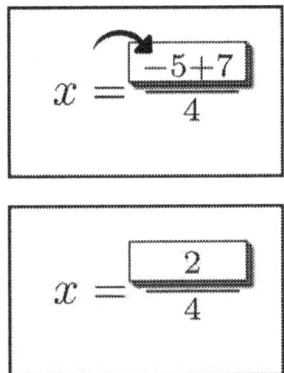

Now that we've reduced everything within the subimages, we continue with the rest of the equation outside of them. Do this by bringing parts of the equation into each subimage, then simplifying. Viewing the equation like this,

$$x = \frac{\boxed{-12}_{2}}{4}$$

bring the $\frac{}{4}$ into each subimage.

$$x = \boxed{\frac{-12}{4}}_{\frac{2}{4}}$$

You can do this with each subimage, one at a time, or do it with both at once. Next, just simplify within each card, starting with the top one,

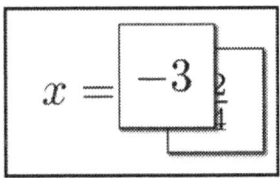

then swapping to the other one and simplifying it.

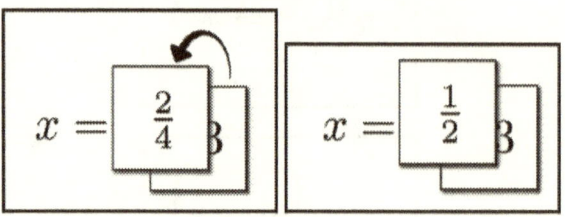

You get the two solutions, $x = \frac{1}{2}$ and $x = -3$.

In this example, we are already pretty close to the solution when the split occurs and we start using subimages. With more complex situations, we will have more manipulations on the subimages themselves; here, we just had one. In larger equations, when it is time to bring part of the equation into a subimage, it is normally best to do that operation on all the subimages before moving on to the next operation. This may be confusing, so here is an example. Say we are solving this equation:

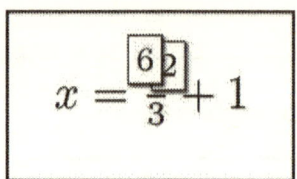

Here, we have had a split, and have already created two subimages. (That's a 6 in the top subimage and a 2 in the bottom one.) To get the pair of final solutions, we must do two things to each value: divide them by 3, then add 1 to the result. There are two approaches you can take to doing that. One path – the path that is generally recommended – is to first bring the $\frac{}{3}$ into each subimage:

$$x = \boxed{\frac{6}{3}\frac{2}{3}} + 1$$

becomes

$$x = \boxed{2\tfrac{2}{3}} + 1$$

After doing this with both subimages, then bring the +1 into each subimage.

$$x = \boxed{\dfrac{2+1}{3}} + 1$$

becomes

$$x = \boxed{3\tfrac{5}{3}}$$

So like I said, all of that – bringing more and more of the surrounding equation into the subimages, one "step" at a time – is one path you can take, and the path that is generally recommended. The other path is to bring the whole equation into each subimage at once.

$$x = \boxed{\dfrac{2}{3}} \boxed{\dfrac{6}{3}+1}$$

This is essentially envisioning a multi-image. With an equation as simple as this one, it is relatively easy to do this. With bigger equations, it usually not so easy, and subimages will be very handy.

At first, when solving plural equations, you will probably want to bring one operation at a time into the subimages. As you get better, you will be able to bring in several at a time. The more operations (steps) you can bring into the subimages at a time, the faster you will get to the solution, provided you bite off what you can chew.

6.3   Exercises

1. Solve these equations using multi-images.

   (a) $(y + 2)^2 = 9$

   (b) $(x - 1)^2 = 4$

   (c) $(x + 1)^2 + 3 = 7$

2. Visualize each of these expressions using subimages. Focus on each version, one at a time, flipping through the different subimages.

   (a) $3 \pm x$

   (b) $\frac{y+4}{x\pm1}$

   (c) $\frac{2x\pm3}{y\pm2}$ (You will flip through four different combinations.)

3. Solve each of these equations by using subimages.

   (a) $-x^2 + 5x - 4 = 0$

   (b) $(x + 2)^2 + 3 = 7$

   (c) $cos\pi(x - 2) = .5, 1 \leq x \leq 3$

Solutions:

1(a) $y = -5$ and $y = 1$ (b) $x = -1$ and $x = 3$ (c) $x = 1$ and $x = -3$

3(a)$x = 1$ and $x = 4$ (b) $x = 0$ and $x = -4$ (c) $x = \frac{5}{3}$ and $x = \frac{7}{3}$

Chapter 7

Epilogue

There is a lot of jargon introduced in this book: symbol motion, chunking, fusion, morphing, subimages, and so on. As far as I know, all of them are new. I wrote this book by carefully examining what I do inside when I solve equations. I did my best to put in writing how it happens, using language you are already familiar with. Sometimes I came across concepts that just didn't seem to already have labels neatly defining them. I think each of them have probably been defined somewhere, by someone. But I haven't found them in the context they are in this book. Even if I had, these concepts are not (yet!) well known. In these cases, I invented terms to describe them, and gave them (hopefully) useful definitions.

Perhaps as a result of how they were "invented", the divisions among these terms are not sharply defined. Symbol motion and fusion do seem to be pretty different concepts, and once you are familiar with them you will never confuse the two. But it is not a great leap to see each of those as a kind of morph. Morphing as described in chapter 5 is something you apply to an expression or a small group of symbols, but there is also a more general sense of the word. The general sense implies a systematic change in the structure of an image. In that larger sense, a symbol motion is just a particular kind of morph of the whole equation. Fusion is a morph of an image of an expression into another expression that takes less ink to write. You can't get *too* loose with this. The mathematical logic must always be preserved, no matter what, or we forsake the whole point. But assuming we are smart enough to stay grounded there, we could summarize everything in this book with the sentence "You can solve equations by visualizing them and morphing that image". We introduce symbol motion, chunking, and so on because they are sub-patterns of morphing that are commonly useful and relatively easy to learn. It's good to occasionally remember that these are all simplifications of what is going on.

You can invent your own patterns. If you practice the techniques in this book, and pay attention to what you are doing and how, you will start to notice subtle things about how your mind works. You may notice that you can do something, but aren't sure how you do it. Maybe you can factor certain polynomials easily, for example, and you usually think nothing of it because it seems normal to you. Then one day you may think about it, and wonder *how* you can do that, and surprisingly you won't immediately be able to say. If you keep looking at it, you may eventually learn how. Maybe it's something you can articulate and maybe it's not. Sometimes things that happen inside of you are like that. It will be known to you, though, and it may open your awareness to how you do some things mathematically. These will be things you already knew, you just *forgot* you knew them.

If you work with equations in a way that you find yourself executing a particular procedure over and over, it will start to become more automatic. It could be almost anything that is a repeated pattern. Maybe you do a particular sort of symbol motion, followed by a fusion of the chunk and the symbol beside it, followed by a morph. Or it could be something much simpler. Whatever, you may find that you are using this pattern in many different equations and different situations. At some point, instead of seeing it as a series of operations, you will sense it as one "thing". It will be one mental operation that you apply to the situation in one fell swoop. That is exactly how I came up with symbol motion. I just observed what I was doing, and noticed that I was doing something similar in many different circumstances. I formalized it a bit, slapped a label called "symbol motion" on the concept, and wrote a chapter in this book detailing it. You can do the same thing. Maybe it will be something that others can use, that you can describe and communicate to people in a way that they can adopt it. Or maybe it is something individual to you, that you find valuable but others won't really find useful because they are not you. It is kind of interesting how it works out.

Everything in this book is just a suggestion. In the community of people who write software using the Perl programming language, there is an expression, TMTOWTDI, pronounced "tim toady". It stands for "There's More Than One Way To Do It." They say that because the language is designed so that there are often several different, equally valid ways to code an algorithm. The idea is that you do it the way that works for you in that situation. The same applies here, only to a greater degree, because what goes on between your ears is much more expressive than something like a language can be. Who decides what the "correct" way is? You do. If you are following some of the instructions in this book, and one day you notice that there is another way to do something, who decides if it is all right to do that? You do. This may seem a weird way to learn math. Maybe it is, but you have the ability to do it this way.

Thank you for reading this book. I hope you enjoy using it as much as I enjoyed writing it.

Exercise

Find something you can do mathematically and which most people cannot do. Teach someone how to do it.

Glossary

chunk (as a verb) Put an expression or collection of symbols together in a group, so that something can be done mathematically with it.

chunk (as a noun) An expression or collection of symbols that are put into a group, so that something can be done mathematically with it.

equation A mathematical statement that two expressions are equal to each other. Generally, anything with an equals sign ($=$) in it.

expression A collection of quantities, variables and operations. It evaluates to a number, if specific values are plugged in for the variables. $2 + x$, 3, and $\frac{x^2+1}{y^2-1}$ are all expressions.

fission A kind of morph in which a group of symbols is replaced with a larger group of symbols. Both groups must be mathematically equal to each other.

fusion A kind of morph in which a group of symbols is replaced with a smaller or simpler group of symbols. Both groups must be mathematically equal to each other.

morph A visual change in which an expression is replaced with a different expression. Both expressions must be mathematically equal to each other. Normally replacing the expression moves the equation closer to the solution.

multi-image Used when solving plural equations. A multi-image is an image containing two or more whole equations, each coming from the same plural equation before its split. Each equation in the multi-image will lead to a solution of the equation.

operation A symbol that signifies two expressions are being combined in some way to form another expression. $+$, $-$, $\times$ and $\div$ are operations.

plural Describes an equation with multiple solutions, and which - at some point during the process of solving it - cannot be represented by a single equation (or simple image of an equation).

quantity A number, variable, or expression of numbers and variables that can evaluate to a number.

relation A symbol that signifies some quantitative relationship between two or more expressions. While the only relation discussed in this book is the equals sign ($=$), other examples are inequalities such as $<$ or $>$.

singular Describes an equation that can be solved by modifying an image of a single equation until it arrives at the solution.

split A split is a point in the process of solving a plural equation in which it cannot be further solved using a single image, because multiple solutions have surfaced. Further processing (solving) requires the use of devices such as multi-images or subimages.

subimage Used when solving plural equations. After a split in a plural equation, often the equations for each solution are similar except for a small sub-expression. A subimage is a portion of the larger image of the equation, usually containing only the sub-expression that is different. There will be two or more subimages, each containing a different sub-expression, and leading to a different solution of the equation.

symbol A character or glyph that has some mathematical meaning. It can be a quantity, variable, relation, or operation.

symbol motion A change in an image representing an equation or expression, which has a visual logic and produces a mathematical change in what is represented by the image. In particular, a symbol motion involves the movement or rearrangement of one or more symbols in the equation, with minor or no alterations. By definition, symbol motions honor the mathematical logic of the transformation. If some pattern does not preserve the mathematical logic, then it is not symbol motion.

visual logic The systematic pattern of change in an image. Usually it means an image of an equation, and that the symbols in the image are moved or changed in a particular, abstract way.

About The Author

Aaron Maxwell has a B.S. in physics from the University of Texas at Dallas. He has had graduate-level training in biophysics and neuroscience, and has worked as a software developer. He is the founder and owner of Hilomath (`http://hilomath.com`), a mathematics education company. Currently he lives in San Francisco, California, USA, where he is writing *Inner Math: Mental Skills for Algebra, Calculus, and More*. He can be contacted by email at amax@hilomath.com.

www.ingramcontent.com/pod-product-compliance
Lightning Source LLC
Chambersburg PA
CBHW022112170526
45157CB00004B/1591